PROFITABLE SHEEP FARMING

Key Strategies For A Thriving Livestock
Business: A Step-By-Step Guide From
Breeding To Successful Ventures

PRESTON DAMIEN

Contents

CHAPTER ONE 8

An Introduction To Sheep Farming 8

Why Should You Consider Investing In Sheep
Farming? 8

Sheep Farming: A Brief History 9

The Modern Importance Of Sheep Farming 9

Types Of Sheep....................................10

Sheep Farming Business Opportunities..........10

How To Get Started With Sheep Farming.......11

Establishing Farm Objectives And Goals........12

Law And Regulation Considerations..............12

Choosing The Best Location 13

Farm Infrastructure And Layout....................13

Buying Sheep Vs. Buying Livestock................14

CHAPTER TWO....................................16

Breeding...16

Sheep Breeding Groups 16

Popular Meat Breeds.............................16

Animals That Produce Wool 17

Breeds That Fulfill Two Purposes 17

Breeds That Are Uncommon Or Historic 18

Sheep Nutrition And Feeding.......................19

Pasture And Grazing Administration...........19

Nutritional Supplements 20

Avoiding Common Feeding Errors...............20

Seasonal Aspects To Consider....................20

CHAPTER THREE 22

Disease And Health Management..................22

Preventive Healthcare Practices 22

Diseases Of Sheep And Their Treatment 23

Vaccination Schedules: 23

Veterinary Attention And Documentation: .23

Reproduction And Breeding Of Sheep............24

System Of Natural Vs. Artificial Breeding 25

Estrus Cycle And Breeding Season 25

Ram Selection And Management....................26

Ewe Management And Pregnancy Care 26

Lamb Management And Care 26

Lamb Management And Rearing....................27

Lamb Nursing For Newborns............................27

Procedures For Weaning.............................28

Growth And Development Stages.................28

Fencing And Lamb Housing...........................29

CHAPTER FOUR...30

Lamb Nutrition For Maximum Growth...........30

Wool Production And Management 30

Methods Of Shearing....................................31

Wool Grading And Marketing 32

Wool Handling And Storage..........................32

Wool Items With Added Value......................33

Marketing And Sales Of Sheep Farming 33

Marketing Strategies For Sheep....................34

Sale Of Live Sheep 35

Offering Lamb Meat 35

CHAPTER FIVE...36

Wool And Wool Products On The Market 36

Customer Relationship Growth.....................36

Sheep Farming Financial Management 37

Budgeting And Financial Planning.................37

Income And Expense Tracking 38

Profitability Analysis..............................38

Managing Hazards In Sheep Farming 39

Environmental And Sustainability Issues In
Sheep Farming...40

Conservation & Biodiversity 41

CHAPTER SIX...44

Sheep Farming Waste Management 44

Future Sheep Farming Trends 46

Technological Progress 47

Emerging Markets (Em) 48

Opportunities And Difficulties 48

Sheep Farming's Role In A Sustainable Future48

Conclusion On Sheep Farming 49

DISCLAIMER

The information provided in this book is intended for general informational and educational purposes only. The author of this book is not engaged in rendering professional or veterinary advice. The content in this book is based on the author's personal experiences, research, and knowledge, and it is not a substitute for professional advice.

The practices and techniques described in this book are meant to provide a foundation for understanding, but specific circumstances may require individualized approaches. It is recommended that readers consult with experienced veterinarians for guidance.

The author and publisher do not guarantee the accuracy, completeness, or timeliness of the information presented in this book. The reader is responsible for ensuring that the practices and recommendations align with current laws, regulations, and safety standards in their respective locations.

By reading this book, you agree that the author will not be held responsible for any actions you take based on the information presented in this book. You are solely responsible for any consequences or outcomes resulting from your farming endeavors. Always exercise caution and seek professional advice when necessary to ensure the well-being of your farm animals and to comply with all relevant laws and regulations.

CHAPTER ONE

An Introduction To Sheep Farming

Sheep farming, also known as ovine husbandry, is an important agricultural industry that includes raising and breeding sheep for several purposes such as wool, meat, and milk. These gentle ruminants have been domesticated for thousands of years and play a vital role in the global agricultural landscape. Sheep farming is practiced over the globe for a range of economic and cultural reasons.

Why Should You Consider Investing In Sheep Farming?

There are many compelling reasons to begin a sheep farm. To begin, sheep are a versatile animal that provides valuable commodities such as wool for textiles, meat for food, and milk for dairy goods. Furthermore, they are adaptable to a wide range of temperatures and may thrive in a variety of environments.

Sheep farming is less costly than other livestock businesses, making it suitable for small and large-scale farmers alike. Furthermore, by restricting weed growth, their grazing actions may assist in creating healthy pastures and aid in sustainable land management approaches.

Sheep Farming: A Brief History

Sheep husbandry has a lengthy history that can be traced back to ancient civilizations like Mesopotamia and Egypt. Sheep have traditionally been prized for their wool, which was vital in the Middle Ages textile industry. The introduction of sheep to new continents by European colonization influenced agricultural systems all across the globe.

The Modern Importance Of Sheep Farming

Sheep herding is still significant in today's world. Wool is still utilized in textile production, while lamb and mutton are popular meals.

Furthermore, because of its role in sustainable agriculture and biodiversity protection, sheep farming has gained ecological value.

Types Of Sheep

Sheep ranching offers a varied selection of breeds, each with unique characteristics. Merino sheep, for example, are recognized for producing high-quality wool, whereas Suffolk animals are favored for meat production. Understanding the differences between breeds is essential for effective sheep herding.

Sheep Farming Business Opportunities

Sheep husbandry provides several economic opportunities. Farmers may choose to specialize in wool production, meat production, or both. Sheep's milk may be used to manufacture high-value products like cheese and yogurt. Ecotourism and educational initiatives centered on sheep herding may also provide income. Furthermore, there is a growth in demand for sustainably bred sheep products, providing opportunities for environmentally conscious businesses to thrive in the market.

Overall, sheep farming offers a diverse set of agricultural techniques for profitability and long-term viability.

How To Get Started With Sheep Farming

Anyone interested in agriculture might consider starting a sheep farm. To begin, would-be farmers should assess their dedication, funds, and expertise.

It is vital to do a study on various sheep breeds and their suitability for your locality. Obtaining required equipment, such as fencing and shelters, is vital for the well-being of the sheep.

It is also crucial to decide if you want a modest hobby farm or a bigger commercial company. Financial success in sheep farming requires a solid business plan and budget.

Establishing Farm Objectives And Goals

A well-defined set of goals and objectives is critical for a profitable sheep farming enterprise. Begin by determining your aim, which might be meat production, wool production, or both. Determine the size of your enterprise and the number of sheep you wish to raise. Set financial goals that take into consideration expenses such as feed, healthcare, and infrastructure. Consider your long-term objectives, such as increasing your flock size or expanding your offers. Goals can assist you in guiding your decision-making and analyzing your progress in the sheep farming business.

Law And Regulation Considerations

Before launching a sheep farming business, it is necessary to navigate the legal and regulatory framework. Understanding zoning constraints, obtaining necessary permissions, and following animal welfare regulations are all part of this.

Registration of your farm and livestock is often needed, and you should be aware of your tax obligations. Also, learn about the local agricultural extension services and support programs that may be available to sheep farmers in your area. Ensure legal compliance from the start to prevent complications later on.

Choosing The Best Location

It is vital to choose the optimal location for your sheep farm. Consider factors such as the climate, soil condition, and market proximity. Adequate access to clean water and grazing areas is crucial for sheep health. Take into account the availability of infrastructure such as roads and electricity. A well-chosen site may have a significant impact on the profitability and sustainability of your sheep farming operation.

Farm Infrastructure And Layout

Designing an appropriate farm plan and infrastructure is crucial for the well-being of your sheep and the total productivity of your farm. Make sure you have enough fences and shelters to protect your flock from severe weather and predators. There should be enough feed and equipment storage facilities. Efficient handling and sorting facilities may aid in speeding up everyday tasks.

A well-planned infrastructure not only preserves your sheep's comfort but also makes management simpler.

Buying Sheep Vs. Buying Livestock

Choosing the right sheep for your farm is a big choice. Take into account the breed, age, and function (meat, wool, or dual-purpose). Examine the animals' health and genetics before purchasing them. Developing relationships with reputable breeders or livestock auctions can help you find high-quality sheep.

Check that you have the necessary facilities and resources to care for your new livestock.

Quarantine and health inspections are essential to prevent the spread of infections on your farm.

A well-planned sheep acquisition strategy establishes the framework for a lucrative and flourishing sheep farming business.

CHAPTER TWO

Breeding

Sheep farming is a centuries-old agricultural practice that includes raising sheep for several purposes such as meat, wool, and even milk. Understanding the many sheep breeds and their characteristics is an essential aspect of sheep care. Sheep breeds differ widely in terms of size, color, coat type, and general suitability for certain jobs.

Sheep Breeding Groups

Sheep breeds are often classified based on their primary purpose: meat production, wool production, dual-purpose, and rare or heritage breeds.

Popular Meat Breeds

Popular meat breeds like the Suffolk and Hampshire are known for their muscular physique and exceptional meat quality. These sheep are good for meat production due to their high muscle-to-fat ratio. They are famous among commercial sheep breeders because of their rapid growth and ability to thrive in a range of situations.

Animals That Produce Wool

Merino and Rambouillet wool-producing breeds are appreciated for their delicate, thick fleeces. These sheep have been purposefully bred for years to produce high-quality wool, which is used in textiles, clothing, and a range of industrial applications. They are often made of soft, crimped fibers that provide superb insulation.

Breeds That Fulfill Two Purposes

Dual-purpose breeds like the Romney and Corriedale produce both meat and wool. They produce a small amount of meat as well as wool, making them ideal for farmers looking to diversify their operations.

Dual-purpose breeds are typically ideal for small-scale, ecologically friendly agricultural operations.

Breeds That Are Uncommon Or Historic

Rare and heritage breeds are rare and may have historical or cultural significance. Examples include Navajo-Churro, Icelandic, and Jacob sheep. These breeds are commonly raised by dedicated conservationists and small-scale farmers who seek to maintain genetic diversity in the sheep population. They may be differentiated by distinguishing characteristics such as multi-horned sheep or certain wool types.

Finally, knowing how to classify sheep breeds is essential for effective sheep herding. The breed selected should be suitable for the farmer's goals, which might be meat, wool, or a mix of the two. Each breed brings unique traits to the farm, and selecting the right one may have a significant impact on the performance and sustainability of the sheep operation. Whether you're a commercial farmer or a follower of heritage breeds, there's something in the world of sheep breeds for everyone in the agricultural business.

Sheep Nutrition And Feeding

Sheep nutrition and feeding are critical components of sheep husbandry success. Providing enough nutrition ensures the health, growth, and productivity of your flock. Sheep are ruminants, which mean their stomachs are divided into four compartments, requiring a specific diet.

Understanding your sheep's nutritional needs is crucial. They need a well-balanced diet rich in carbohydrates, proteins, vitamins, and minerals. Grass and pasture are essential components of their diet because they provide fiber and essential elements. However, to adequately meet the nutritional demands of your sheep, you must assess the nutritional content of your pasture.

Pasture And Grazing Administration

Pasture management is essential in sheep farming. It is vital to keep your pastures in excellent condition if you want to feed your flock. Overgrazing is avoided through rotational grazing, enabling grasslands to recover and keep nutritional value. Weed control, fertilizer, and soil testing are all part of good pasture management.

Nutritional Supplements

Supplemental feeding is essential when pasture quality is poor or during particular life stages, such as pregnancy or lactation. Supplements containing a range of nutrients assist sheep in meeting their dietary requirements for growth and reproduction.

Avoiding Common Feeding Errors

Overfeeding and underfeeding your sheep may both be harmful to their health. Overfeeding may result in obesity and other problems, whilst underfeeding can result in famine. Maintain continual access to clean water and keep an eye on your flock's physical well-being.

Seasonal Aspects To Consider

Seasonal variations have an impact on sheep nutrition. Because forage availability is poor throughout the winter, you may need to increase supplementary feeding.

In the summer, abundant meadows may provide ample nutrition. Adjust feeding operations as required, taking temperature, humidity, and other environmental elements that affect sheep comfort and health into consideration.

The ability to regulate sheep nutrition and feeding needs throughout the year is crucial to the success of sheep farming.

CHAPTER THREE

Disease And Health Management

It is vital in sheep husbandry to keep your sheep healthy. Healthy sheep are not only more productive, but they also reduce the probability of illness outbreaks in your flock. It is vital to regularly check their health.

This includes monitoring their behavior, checking for signs of sickness, and ensuring they have enough nutrition and hygiene. Adequate protection from severe weather is also required.

Preventive Healthcare Practices

Illness prevention is critical to reduce illness incidence in your flock. Implement quarantine for new arrivals, rotational grazing to limit parasite buildup, and proper nutrition to boost the immune system.

Regular hoof trimming and dental treatment are required. Additionally, maintaining clean and dry living conditions assists in the avoidance of common ailments such as foot rot and respiratory infections.

Diseases Of Sheep And Their Treatment

Sheep are susceptible to diseases such as foot rot, pneumonia, and intestinal parasites. Early symptom identification is crucial for successful treatment.

Consult a veterinarian for an accurate diagnosis and treatment plan. Use biosecurity measures to prevent disease from spreading inside your herd.

Vaccination Schedules:

Create a vaccination schedule with the assistance of a veterinarian. Vaccines against Clostridial illness and respiratory infections are critical. Vaccinations given on time may prevent your sheep from potentially fatal illnesses.

Veterinary Attention And Documentation:

Regular veterinary exams are essential. Keep thorough records of all immunizations, treatments, and any health-related issues. By documenting the health history of each sheep, this information aids in disease management approaches.

A proactive approach to sheep farming health and disease prevention supports the well-being of your flock and contributes to a lucrative and sustainable operation.

Reproduction And Breeding Of Sheep

Reproduction and breeding are critical aspects of sheep management because they have a significant influence on flock health and productivity. Proper regulation of these processes ensures a regular supply of lambs for meat, wool, and other items. Sheep have a regular reproductive cycle and typically give birth once a year.

Farmers must frequently monitor the ewes for signs of estrus, which signal the start of the fruitful season.

System Of Natural Vs. Artificial Breeding

In sheep husbandry, the two primary breeding processes are natural and artificial insemination. Natural breeding involves introducing a ram to the ewes and allowing them to mate naturally. This method takes advantage of the ram's ability to distinguish ewes in estrus. In contrast, artificial insemination involves extracting and injecting sperm from a high-quality ram into the ewes. While natural breeding is less costly and simpler, artificial insemination allows for more precise genetic selection and disease control.

Estrus Cycle And Breeding Season

Understanding the sheep's estrus cycle is crucial for effective breeding. Ewes cycles every 17-19 days, with each cycle lasting around 30 hours. The breeding season, when ewes are most receptive, varies per breed, although it is often in the autumn or early winter. Coordination of ewes' estrus cycles is required for successful regulation of reproductive efficiency.

Ram Selection And Management

Selecting healthy and genetically superior rams is crucial for optimal breeding. Hereditary deformities and diseases in rams should be prevented. They need regular nutrition and health care to maintain their fertility and performance. It is also vital to keep an eye on their physical condition and to offer enough shelter and space.

Ewe Management And Pregnancy Care

Ewes need special care throughout pregnancy to ensure the health of both the mother and the maturing lamb. A healthy diet, vaccinations, and regular health checkups are all essential. Ewes should be provided with a clean and comfortable lambing environment as their due date approaches.

Lamb Management And Care

Lambing management includes assisting ewes throughout the birthing process, giving lambs colostrum for immunity, and monitoring their health. Lamb survival needs adequate shelter, nutrition, and predator protection. Early diagnosis and treatment of health issues may significantly improve lambing success and total flock productivity.

Lamb Management And Rearing

Lamb rearing and management are critical components of successful sheep husbandry. The selection of healthy ewes and rams is the first step in the breeding process. Lambs must be properly cared for after birth.

Housing, nutrition, and healthcare are all vital. Lambs are vulnerable in their infancy and must be regularly observed.

This entails providing them with colostrum, a nutrient-rich mother's milk, during their first few hours of life.

Lamb Nursing For Newborns

It is tough to care for newborn lambs. It requires monitoring their health, ensuring they connect with their mothers, and keeping them warm and dry.

Consumption of colostrum is vital because it includes essential antibodies and nutrients. Regular health exams help to discover issues early. Comfortable mattress and shelter to keep you warm.

Procedures For Weaning

Weaning is a key period in a lamb's life. It is most common between the ages of 8 and 12 weeks.

Lambs adjust more quickly when they are progressively taken from their moms and fed enough. During this period, sufficient grazing and water access should be provided.

Growth And Development Stages

Lambs go through various phases of development and growth. They include the neonatal, pre-weaning, weaning, and post-weaning stages. It is vital to evaluate their growth, weight gain, and general health at each stage.

Fencing And Lamb Housing

Adequate housing and fencing assure the safety and well-being of lambs. Properly designed pens or barns protect them from severe weather and predators. Fencing should be secure and age-appropriate for each breed and age group.

CHAPTER FOUR

Lamb Nutrition For Maximum Growth

Feeding is an essential part of lamb growth. Nutrition is critical for good development. Lambs are often fed milk or milk substitutes before progressing to a solid diet. A well-balanced diet strong in protein and essential nutrients promotes healthy development.

To conclude, excellent lamb rearing and management processes are required for successful sheep farming operations. A thriving sheep flock requires close attention to births, proper weaning, monitoring developmental stages, and providing enough shelter and nutrition.

Wool Production And Management

Sheep farming is typically valued for the production of wool, which is an important component of the company. The shearing procedure, which involves the removal of the sheep's fleece, is used to get wool. Proper management is required to produce high-quality wool. This includes feeding, housing, and health care for the sheep.

Farmers purposefully choose sheep breeds known for their wool quality and quantity to increase wool production. Regular brushing and pest inspection help to keep the fleece clean. Shearing occurs once a year, mostly in the spring, to ensure the comfort of the sheep throughout the warmer months.

Methods Of Shearing

Shearing demands dexterity and accuracy. The fleece is tenderly removed from the sheep using specialized shearing machines, ensuring their safety and comfort. Experienced shearers utilize sophisticated processes to guarantee fast and clean shearing.

Working from the hindquarters to the head, correct shearing techniques need caution around sensitive areas such as the belly and udder.

Shearing carelessly may result in cuts or damage to the fleece, decreasing its value.

Wool Grading And Marketing

Wool is graded after shearing depending on factors such as fiber length, diameter, and cleanliness. This grading has an impact on the quality and market value of the wool. Wool is then marketed to industries such as textiles, fashion, and manufacturing.

Wool marketing strategies include selling directly to manufacturers, participating in wool auctions, and cooperating with wool cooperatives. Wool's exceptional qualities, including warmth, durability, and sustainability, are regularly emphasized in marketing efforts.

Wool Handling And Storage

Wool must be treated and kept correctly to keep its quality. Following shearing, the wool is methodically rolled, weighed, and placed in bundles for transport and storage.

It is vital to keep the wool clean and free of pests and moisture.

Wool Items With Added Value

Aside from raw wool, the sheep farming industry relies on value-added wool products to exist. Yarn, clothing, blankets, and insulating materials are only a few examples. Cleaning, carding, spinning, and weaving are standard procedures for transforming wool into a variety of goods that add significant value to the basic material.

Value-added products provide new market opportunities for sheep farmers, such as boutique fashion shops, craft stores, and eco-friendly insulation suppliers, helping them diversify their income streams and reach a larger customer base.

Marketing And Sales Of Sheep Farming

Marketing and sales are crucial to the sheep farming industry's profitability and survival. Promoting and selling a range of sheep-related commodities, such as live sheep, lamb meat, wool, and wool products, is an example of effective marketing.

Marketing Strategies For Sheep

1. Targeted Advertising: Determine your target demographic, whether it's local customers, restaurants, or wholesale purchasers, and adapt your marketing efforts to them. Use social media, websites, and local advertising to reach out to prospective consumers.

2. Product differentiation: To make your sheep goods stand out in the market, emphasize their distinctive attributes, such as grass-fed or organic lamb meat. Highlight your items' health advantages and great quality.

3. Local Collaborations: Build a steady consumer base by collaborating with local eateries and marketplaces. Offering these partners unique prices or exclusive items might help you develop a better market presence.

4. Online Sales: Establish your e-commerce presence to reach a broader audience.

Provide clear product descriptions, high-quality photos, and simple shipping alternatives to online shoppers.

Sale Of Live Sheep

When selling live sheep, consider aspects such as breed, age, and health. Attend local livestock auctions, advertise in agricultural periodicals, or utilize online livestock markets to connect with possible buyers.

Offering Lamb Meat

When selling lamb meat, prioritize quality and appearance. Collaborate with local butchers, restaurants, or farmers' markets to advertise your goods. Offering sample tastes may captivate clients and develop confidence in the quality of your meat.

CHAPTER FIVE

Wool And Wool Products On The Market

Wool items should be promoted since they are warm, durable, and long-lasting. Participate in craft shows, work with local craftsmen, and sell wool and woolen products on internet markets. Highlight the natural and eco-friendly aspects of your items.

Customer Relationship Growth

Customer retention requires long-term customer connections. Provide outstanding customer service, respond quickly to queries, and handle problems as soon as possible. Consider using loyalty programs or newsletters to keep clients involved and up to date on your sheep farming operations.

Engage your audience on social media channels to build community and trust in your business. Finally, a solid client base is the cornerstone of a successful sheep farming firm.

Sheep Farming Financial Management

Financial management is an important element of sheep farming success. Effective financial management maintains the farm's economic sustainability and profitability. Feeding, healthcare, infrastructure, and labor are all costs associated with sheep farming. Farmers must emphasize planning, spending monitoring, profitability analysis, and maybe seek government support and subsidies to guarantee long-term success.

Budgeting And Financial Planning

The cornerstone of financial planning in sheep farming is a well-structured budget. Farmers should estimate their yearly revenue, which should include sales of lambs, wool, and other byproducts. Feed, veterinary care, equipment upkeep, and labor expenditures must all be considered.

A budget assists farmers in allocating resources effectively and avoiding overspending, providing financial stability for the farm.

Income And Expense Tracking

Maintaining accurate records of income and expenses are critical for financial management. Farmers should keep meticulous records of all dealings. They may use this data to find areas where expenses can be cut and revenue can be enhanced. This process may be sped up by using digital tools and software, which make it simpler to examine financial data and make educated choices.

Profitability Analysis

Profitability analysis entails evaluating the farm's financial performance over time. Farmers should assess the return on investment (ROI) for several elements of their business, such as breeds, feeding tactics, and healthcare procedures. This study assists farmers in making data-driven choices to enhance revenue.

Government aid and Grants Many governments provide aid and grants to agricultural industries, including sheep farming. Farmers can investigate these options for obtaining finance for infrastructure upgrades, R&D, or environmental activities. Grants and subsidies may greatly increase a farm's financial resources.

Managing Hazards In Sheep Farming

Disease outbreaks, market changes, and poor weather conditions are all hazards in sheep farming. Risk management solutions that are effective are critical for financial stability. Farmers may reduce risks by diversifying their revenue streams, adopting biosecurity measures, and setting aside emergency cash to meet unforeseen expenditures.

Finally, financial management is critical to the success of sheep farming. Farmers may assure the economic sustainability and long-term profitability of their enterprises by emphasizing budgeting, revenue and spending monitoring, profitability analysis, pursuing government aid, and risk management. Proper financial management helps not just individual farms, but also the entire health and viability of the sheep farming business.

Environmental And Sustainability Issues In Sheep Farming

Sheep farming is important in agriculture, however, it is critical to use sustainable ways to reduce its environmental effect. Sustainable sheep farming requires a multifaceted strategy that takes into account ecological, economic, and social concerns.

Implementing Sustainable Farming Practices in Sheep Farming entails maximizing land use, decreasing water and chemical consumption, and enhancing animal welfare. Rotational grazing is a sustainable approach for improving pasture health and reducing soil erosion. Selective breeding for disease resistance and enhanced genetics may increase flock health while reducing antibiotic use.

Conservation & Biodiversity

Sheep husbandry and biodiversity conservation may coexist. Local species may find sanctuary on farms by maintaining hedgerows, growing native plants, and protecting natural habitats. Farmers may help safeguard endangered species and sustain healthy ecosystems by avoiding overgrazing and implementing prudent land management.

Waste Management: Waste management is critical for long-term sustainability. Sheep husbandry creates organic waste, which, when handled properly, may be converted into valuable compost or utilized to collect methane. This lowers greenhouse gas emissions and promotes circular agriculture.

Energy Efficiency: By using energy-efficient technology such as solar-powered water pumps, LED lighting, and efficient heating systems, sheep farming may lower its carbon impact. Furthermore, minimizing needless transit and streamlining supply chains might help to reduce energy use even further.

Finally, sustainable sheep farming requires a holistic strategy that emphasizes environmental stewardship, biodiversity protection, waste reduction, and energy efficiency.

Sheep producers may maintain the long-term success of their enterprises while limiting their environmental effects by embracing these strategies.

CHAPTER SIX

Sheep Farming Waste Management

Waste management is a vital component of good sheep husbandry. Sheep husbandry creates organic waste, such as manure and bedding materials, which may have negative environmental consequences if not effectively handled.

Here are some important factors to consider when it comes to waste management in sheep farming:

Manure Management: Sheep create a lot of manure, which, if not handled properly, may pollute the water and cause nutritional imbalances in the soil. Farmers may ensure proper manure management by composting, spreading manure uniformly, and minimizing runoff into water bodies. Composting not only eliminates trash but also produces nutrient-rich organic matter that may be utilized to boost soil fertility.

Bedding Materials: It is important to properly dispose of old bedding materials like straw or hay. These products may be recycled or used to decrease waste and save money. Used bedding, for example, may be composted with manure to generate an excellent soil conditioner.

Sheep dung is a source of methane emissions, a powerful greenhouse gas.

Some farms are using anaerobic digestion systems to absorb methane from manure and convert it into biogas for electricity production. This method not only eliminates trash but also adds to the creation of renewable energy.

Regulatory Compliance: It is essential to follow local waste management rules. To prevent environmental breaches, farms must be aware of and follow rules governing manure storage, disposal, and nutrient management programs.

Implementing sustainable agricultural methods such as rotational grazing and reducing the use of chemical fertilizers may lead to improved nutrient management and waste reduction.

Finally, waste management is critical for environmental sustainability in sheep production. Responsible waste disposal, composting, methane capture, and regulatory compliance are critical components of sustainable waste management on sheep farms. Sheep producers may lessen their environmental impact and contribute to a more sustainable agriculture economy by using these strategies.

Future Sheep Farming Trends

Sheep farming has a promising future as it adjusts to changing customer demands and global issues. One notable trend is the growing demand for items made from ethically reared lamb and wool. Farmers are becoming more ecologically sensitive as a result of this change in consumer behavior.

Breeders are also harnessing technology to generate sheep that are more disease-resistant, produce larger yields, and have better meat quality. Automation and precision agricultural methods are cutting labor expenses and increasing overall efficiency.

Technological Progress

Sheep farming is going through a technological revolution. Precision agricultural techniques such as GPS trackers and automatic feeding allow farmers to watch their flocks with unparalleled precision.

This data-driven approach to flock management improves flock management, resulting in healthier sheep and higher output.

Genetic advances are producing sheep breeds that are more disease-resistant and generate higher-quality meat and fiber. Genetic editing tools are set to significantly transform breeding programs by allowing farmers to adapt their flocks to particular market needs.

Emerging Markets (Em)

The world's sheep farming sector is growing into new markets. Increased demand for lamb and wool products is being driven by rising middle-class populations in countries such as China and India. This opens up chances for international commerce and industrial partnerships.

Opportunities And Difficulties

While climate change, land use limits, and market volatility provide problems to sheep farming, they also bring possibilities. Sustainable methods may help reduce climate-related hazards while diversifying into value-added items such as specialty cheeses and organic wool can boost profitability.

Sheep Farming's Role In A Sustainable Future

Sheep farming is critical to a sustainable future. These creatures are skilled at transforming marginal areas into lucrative resources, sequestering carbon, and promoting biodiversity. In a constantly changing world, the sector can contribute to food security and environmental stewardship by adopting sustainable practices and technologies.

Conclusion On Sheep Farming

To summarize, sheep farming is an important component of world agriculture, providing several advantages to both farmers and society as a whole. Sheep domestication has a rich history that has developed into a contemporary business distinguished by ecological methods, economic viability, and a varied variety of products.

These animals supply us with wool, meat, and dairy products, which help to ensure our food security and the textile industry.

Sheep farming also helps to preserve ecological balance by grazing on grasslands, avoiding overgrowth, and reducing the danger of wildfires. Furthermore, their excrement improves soil fertility, enabling healthy crop development.

However, the sector has several hurdles, including disease control, market swings, and environmental issues. Sheep farming demands ongoing research and innovation, as well as prudent management, to maintain its long-term viability.

With a rising global population and increased demand for sustainable and locally produced goods, sheep farming is positioned to remain a crucial and versatile agricultural industry, offering significant resources for future generations.

www.ingramcontent.com/pod-product-compliance
Lightning Source LLC
Chambersburg PA
CBHW060007300526
45794CB00003B/1123